A TREATISE

ON

STAFF MAKING

AND

PIVOTING

CONTAINING COMPLETE DIRECTIONS FOR MAKING AND FITTING NEW STAFFS FROM THE RAW MATERIAL.

EUGENE E. HALL

WITH NUMEROUS ILLUSTRATIONS

CHICAGO:
GEO. K. HAZLITT & CO., PUBLISHERS.
1896.

CONTENTS.

STAFF MAKING AND PIVOTING.

CHAPTER I.

TO produce a good balance staff requires more skill than to produce any other turned portion of a watch, and your success will depend not alone on your knowledge of its proper shapes and measurements, nor the tools at your command, but rather upon your skill with the graver and your success in hardening and tempering. There are many points worthy of consideration in the making of a balance staff that are too often neglected. I have seen staffs that were models as regards execution and finish, that were nearly worthless from a practical standpoint, simply because the maker had devoted all his time and energy to the execution of a beautiful piece of lathe work, and had given no thought or study to the form and size of the pivots. On the other hand, one often sees staffs whose pivots are faultless in shape, but the execution and finish so bungling as to offset all the good qualities as regards shape. To have good tools and the right idea is one thing, and to use these tools properly and make a practical demonstration of your theory is another.

I shall endeavor to take up every point in connection with the balance staff, from the steel to the jewels, and their relation to the pivots, and I believe this will then convey to the reader all the necessary points, not only as regards staffs, but pivots also, whether applied to a balance or a pinion staff.

It may be argued, and we often do hear material dealers advance the theory, that to-day, with our interchangeable parts and the cheapness of all material, it is a waste of time to make a balance staff. To the reader who takes this view of the situation I simply want to say, kindly follow me to the end of this paragraph, and if you are still of the same opinion, then you are wasting your time in following me farther. For a material dealer to advance this theory I can find some excuse; he is an interested party, and the selling of material is his bread and butter; but the other fellow, well I never could understand him and possibly never shall. When we seriously consider the various styles and series in "old model" and "new model," of only one of the leading manufacturers of watches in this country, to say nothing of the legion of small and large concerns who are manufacturing or have manufactured in the past, and then think of carrying these staffs in stock, all ready for use, we then begin to realize how utterly absurd the idea is, to say nothing of how expensive! On the other hand, if you reside in a large city and propose to rely on the stock of your material dealer, you will find yourself in an embarrasing situation very often, for as likely as not the movement requiring a new staff was made by a company that went out of business back in the '80s, or it is a new movement, the material for which has not yet been placed on the market. This state of affairs leads to makeshifts, and they in turn lead to botch work. The watchmaker who does not possess the experience or necessary qualifications to make a new balance staff and make it in a neat and workmanlike manner, is never certain of having exactly what is needed, and cannot hope to long retain the confidence of his customers. In fact, he is not a watchmaker at all, but simply an apprentice or student, even though he be working for a salary or be his own master. There are undoubtedly many worthy members of the trade who are not familiar with the making of a balance staff, who will take exceptions to this statement; but it is nevertheless true. They may be good workmen as far as they go; they may be painstaking; but they cannot be classed as watchmakers.

This article is intended for the benefit of that large class whose opportunities for obtaining instruction are limited and who are ready and willing to learn, and for that still larger class of practical workmen who can make a new staff in a creditable manner, but who are always glad to read other people's ideas on any subject connected with the trade and who are not yet too old to learn new tricks should they find any such.

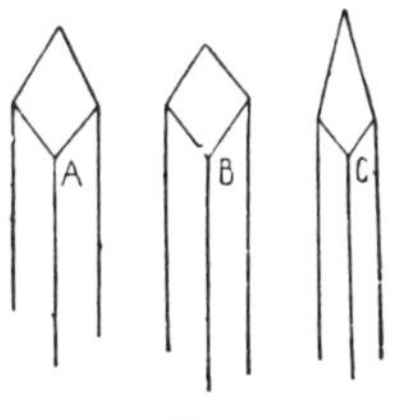

Fig. 1.

Good tools, in good condition, are the most essential requisites in making a new staff. I would not advise any particular make of lathe, as the most expensive lathe in the world will not produce a true staff if the workman cannot center his work accurately and does not know how to handle his graver, while on the other hand fine work can be done on the simplest and cheapest lathe by a workman possessing the requisite skill. I will take it for granted that you use an American-made lathe of some kind, or a foreign-made lathe manufactured on American lines. It is advisable, though not absolutely necessary, to have three gravers similar to those illustrated in Fig. 1, A being used for turning the staff down in the rough; B for the conical pivots and square shoulders and C for the under-cutting. The other tools and attachments needed will be described as I come to them in use.

The balance staff should be made of the best steel, tempered to such a degree as to give the longest service and yet not so hard as to endanger the breakage of the pivots. Select a piece of Stubb's steel wire, say No. 46, or a little larger than the largest part of the finished staff is to be, and center it in a split chuck of your lathe. Be careful in selecting your chuck that you pick one that fits the wire fairly close. The chuck holds the work truest that comes the nearest to fitting it. If you try to use a chuck that is too large or too small for the work, you will only ruin the chuck for truth. Turn the wire to the form of a rough staff, as shown in Fig. 2, leaving on a small part of the original wire, as shown at A. After the wire is roughed out to this general form, remove from the chuck and get ready to harden and temper it. The hardening and tempering may be effected in various ways, and I am scarcely prepared to say which method is the best, as there are several which give about the same general results. One method of hardening is to smear the blank with common yellow soap, heat it to a cherry red, and drop endwise into linseed oil. Petroleum is preferred by some to linseed oil, but, to tell the truth, I can see no difference in the action of linseed, petroleum or olive oil. Be sure and have enough oil to thoroughly cool the blank, and a deep vessel, such as a large-mouthed vial, is preferable to a saucer. The blank will now be found too hard to work easily with the graver, and we must therefore draw the temper down to that of fine spring steel. Before doing this the blank should be brightened, in order that we may see to just what color we are drawing it. The main object in using the soap in hardening is that it may form a scale upon the blank, and if the heating is effected gradually the soap will melt and form a practically air-tight case around the blank. This scale, if the hardening is carefully and properly done, will generally chip and fall off when the blank is plunged in the oil, particularly if the oil is cool, and if it does not fall off of its own accord, it can easily be removed by rolling the blank upon the bench. If it does not come out clean, or if soap is not used, it may be brightened by again inserting in the lathe and bringing it in contact with a piece of fine emery paper or cloth.

Fig. 2.

I draw the temper in the following manner: Place some fine brass filings in a boiling-out cup or bluing pan and lay the blank upon these filings, holding the pan over the flame of an alcohol lamp until the blank assumes a dark purple color, which it will reach when the heat gets to about 500° F. This I consider the right hardness for a balance staff, as it is not too hard to work well under the graver nor too soft for the pivots. At this degree of hardness steel will assume an exquisite polish if properly treated. Another method of tempering is to place the staff on a piece of sheet iron or copper (say 1 inch wide by 4 long), having previously bent it into a small angle for the reception of the staff, as shown in Fig. 3. This piece of metal, when nicely fitted into a file handle, will answer all the purposes of the bluing pan and presents quite a neat appearance. Having placed the blank in the angle, lay on it a piece of yellow wax about the size of a bean, and heat it over your lamp until the wax takes fire and burns. Blow out the flame and allow the staff to cool, and it will be found to be of about the right hardness.

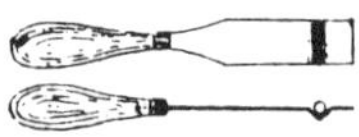

Fig. 3.

We have now arrived at an important station in staff making, a junction, we may term it, where many lines branch off from the main road. At this particular spot is where authorities differ. I have no hesitation in saying that at this particular point the split chuck should be removed from the lathe head and carefully placed in the chuck box and the cement chuck put in its place. I believe that all the remaining work upon a staff should be executed while it is held in a cement chuck. On the other hand I have seen good workmen who turned and finished all the lower part of a staff while in a split chuck, cut it off and turned and finished the upper part in a cement chuck. All I have got to say is that they had more confidence in the truth of their chucks than I have in mine. I have even read of watchmakers who made the entire staff in a split chuck, but I must confess I am somewhat curious to examine a staff made in that way, and must have the privilege of examining it before I will admit that a true staff can be so made.

We will suppose that the workman has a moderately true chuck, and that he prefers to turn and finish all the lower portions in this way. Of course the directions for using a cement chuck on the upper part of a staff are equally applicable to the lower. Before going further I think it advisable to consider the requirements of a pivot, but will reserve this for another chapter.

CHAPTER II.

THE chief requirements of a pivot are that it shall be round and well polished. Avoid the burnish file at all hazards ; it will not leave the pivot round, for the pressure is unequal at various points in the revolution. A pivot that was not perfectly round might act fairly well in a jewel hole that was round, but unfortunately the greater proportion of jewel holes are not as they should be, and we must therefore take every precaution to guard against untrue pivots. Let us examine just what the effect will be if an imperfect pivot is fitted into an unround hole jewel, and to demonstrate its action more clearly let us exaggerate the defects. Suppose we pick a perfectly round jewel and insert into the opening a three-cornered piece of steel wire, in shape somewhat resembling the taper of a triangular file. We find that this triangular piece of steel will turn in the jewel with the same ease that the most perfect cylindrical pivot will. Now suppose we change the jewel for one that is out of round and repeat the experiment. We now find that the triangular steel soon finds the hollow spots in the jewel hole and comes to a stand-still as it is inserted in the hole. The action of a pivot that is not true, when in contact with a jewel whose hole is out of round, is very similar, though in a less marked degree. If the pivot inclines toward the elliptical and the jewel hole has a like failing, which is often the case, it is very evident that this want of truth in both the pivot and hole is very detrimental to the good going of a watch.

There are two kinds of pivots, known respectively as straight and conical pivots, but for the balance staff there is but one kind and that is the conical, which is illustrated in Fig. 4. The conical pivot has at least one advantage over the straight one, *i. e.*, it can be made much smaller than a straight pivot, as it is much stronger in proportion, owing to its shape. All pivots have a tendency to draw the oil away from the jewels, and particularly the conically formed variety, which develops a strong capillary attraction. To prevent this capillary attraction of the oil, the back-slope is formed next to the shoulder, although many persons seem to think that this back-slope is merely added by way of ornament, to make the pivot more graceful in appearance. It is very essential, however, for if too much oil is applied the staff would certainly draw it away if its thickness were not reduced, by means of the back-slope. Before leaving the subject of capillarity let us examine the enlarged jewel in Fig. 5; c is an enlarged pivot, b is the hole jewel and a is the end stone. We observe that the hole jewel on the side towards the end stone is convex. It is so made that through capillarity the

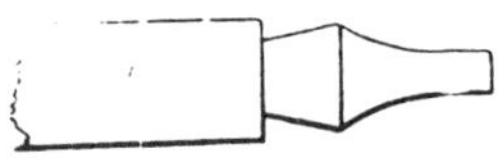

Fig. 4.

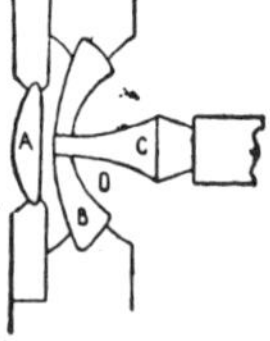

Fig. 5.

oil is retained at the end of the pivot where it is most wanted. It is, in my opinion, very necessary that the young watchmaker should have at least a fair understanding of capillarity, and should understand why the end stone is made convex and the pivot with a back slope. For this reason I will try and make clear this point before proceeding further. We all know that it is essential to apply oil to all surfaces coming in contact, in order to reduce the friction as much as possible, and if the application of oil is necessary to any part of the mechanism of a watch, that part is the pivot. Saunier very aptly puts it thus: "A liquid is subject to the action of three forces: gravity, adhesion (the mutual attraction between the liquid and the substance of the vessel containing it), and cohesion (the attractive force existing among the molecules of the liquid and opposing the subdivision of the mass.)"

We all know that if we place a small drop of oil upon a piece of flat glass or steel and then invert the same the oil will cling to the glass, owing to the adhesion of the particles; if we then add a little more to the drop and again invert, it will still cling, although the drop may be elongated to a certain degree. This is owing to the cohesion of the molecules of the oil, which refuse to be separated from one another. If, however, we again add to the drop of oil and invert the plate the drop will elongate and finally part, one portion dropping while the other portion clings to the main body of the liquid. The fall of the drop is occasioned by gravity overcoming the cohesion of the molecules. Now take a perfectly clean and polished needle and place a drop of oil upon its point and we will see that the oil very rapidly ascends towards the thicker portion of the needle. Now if we heat and hammer out the point of the needle into the form of a small drill and repeat the operation we find that the oil no longer ascends. It rises from the point to the extreme width of the drill portion, but refuses to go beyond. It clings to that portion of the needle which would correspond to the ridge just back of the slope in a conical pivot. Water,

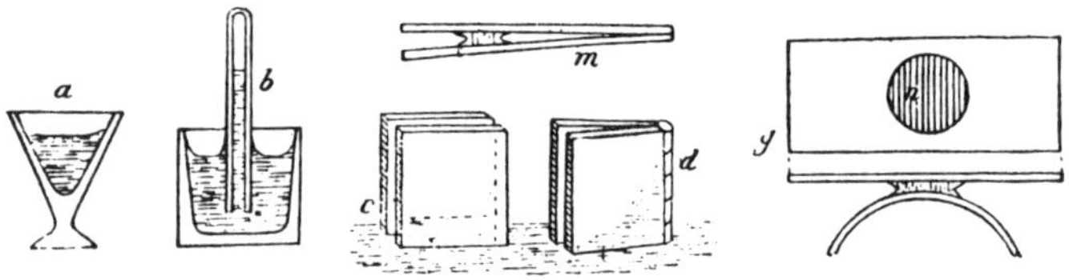

Fig. 6.

oil, etc., when placed in a clean wine glass, do not exhibit a perfect level surface, but raise at the edges as shown at a in Fig. 6. If a tube is now inserted, we find that the liquid not only rises around the outside of the tube and the edges of the vessel, but also rises in the tube far beyond its mean level, as shown at b. These various effects are caused by one of the forces above described, *i. e.*, the adhesion, or mutual attraction existing between the liquid and the substance of the vessel and rod. The word capillarity is of Latin derivation, and signifies hair-like slenderness. The smaller the tube, or the nearer the edges of a vessel are brought together, the higher in proportion will the liquid rise above the level. An ascent of a liquid, due to capillarity, also takes place, where the liquid is placed between two separate bodies, as oil placed between two pieces of flat glass. If the plates are parallel to one another and perpendicular to the surface of the liquid it will ascend to the same height between the plates, as shown at c in Fig 6. If the plates were united at the back like a book and spread somewhat at the front, the oil would ascend the higher as the two sides approach one another, as shown at d, Fig. 6. If a drop is placed somewhat away from the intersecting point of the glasses, as shown at m it will, if not to far away, gradually work its way to the junction, providing the glasses are level. If, however, the glasses are inclined to a certain extent, the drop will remain stationary, since it is drawn in one direction by gravity and in the other by capillarity. When a drop of oil is placed between two surfaces, both of which are convex, or one convex and the other plain, as shown at g, it will collect at the point n, at which the surfaces nearest approach one another. We now see very clearly why the hole jewel is made convex on the side towards the end-stone and concave on the side towards the pivot.

Particular pains should be taken to polish those portions of the pivots which actually enter the jewel hole and to see that all marks of the graver be thoroughly removed, because if any grooves, no matter how small, are left, they act as minute capillary tubes to convey the oil.

If the hole jewel be of the proper shape, the endstone not too far from the hole jewel and too much oil is not applied at one time, the oil will not spread nor run down the staff, but a small portion will be retained at the acting surface of pivot and jewel, and this supply will be gradually fed to these parts from the reservoir between the jewel and end-stone, by the action of capillarity.

Having examined into the requirements of the pivot and its jewel and having gained an insight into what their forms should be, we are the better able to perform that portion of the work in an intelligent manner.

CHAPTER III.

OUR wire has been roughed out into the form of a staff, has been hardened and the temper drawn down to the requisite hardness and we are now ready to proceed with our work. As I said before, we have now arrived at a point where many authorities differ, *i. e.*, as to whether the finishing of the staff proper, should be performed while the work is held in the chuck, or whether a wax chuck be substituted. We will take it for granted that you have a true chuck and that you prefer to finish all the lower portion of the staff while held in the chuck.

Before we proceed with our work it will be necessary for us to make some accurate measurements, as we cannot afford to do any guess work by measuring by means of the old staff. I have used a number of different kinds of calipers and measuring instruments for determining the various measurements for a balance staff, but have met with more success with a very simple little tool which I made myself from drawings and description published some years ago in THE AMERICAN JEWELER. This simple little tool is shown in Fig. 7, and has been of great service to me. It consists of a brass sleeve A, with a projection at one end as shown at B. This sleeve is threaded, and into it is fitted the screw part C, which terminates in a pivot D, which is small enough to enter the smallest jewel. The sleeve I made from a solid piece of brass, turning it down in my lathe and finishing the

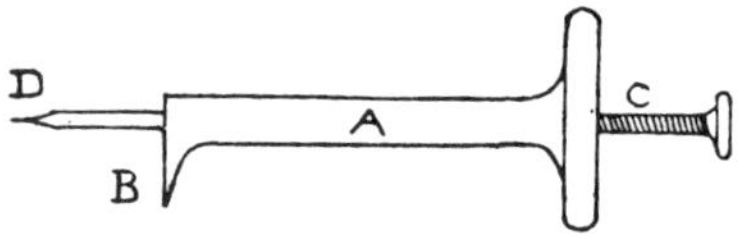

Fig. 7.

projection by means of a file. The hole was then drilled and threaded with a standard thread. The screw part C, I made of steel and polished carefully.

To ascertain the proper height for the roller, place it upon the tool, allowing it to rest upon the leg B, and set the pivot D in the foot jewel. Now adjust, by means of the screw C until the roller is in its proper position in relation to the lever fork. This may be understood better by consulting Fig. 8, where A is the guage, C is the roller, E is the lever, F. is the plate and G is the potance.

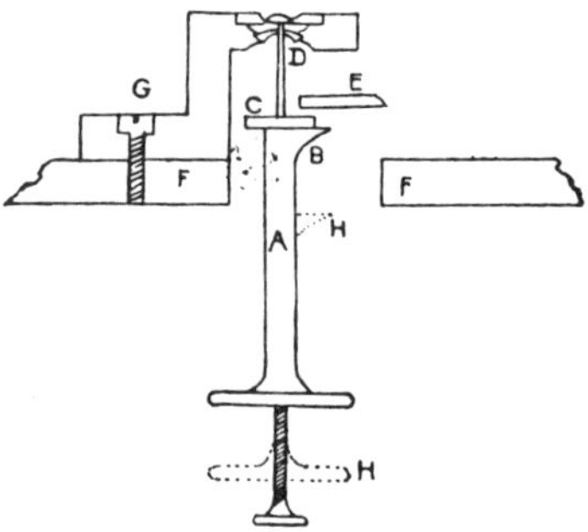

Fig. 8.

Now in order to locate the proper place to cut the seat for the roller, remove it from the foot of the gauge and apply the guage to the work as shown in Fig. 9. The foot of the guage resting against the end of the pivot, the taper end of the gauge will locate accurately the position of the roller seat. In order to locate the proper position for the seat for the balance, proceed the same as for the roller, except that the foot of the gauge is lowered until it is brought sufficiently below the plate to allow of the proper

Fig. 9.

Fig. 10.

clearance as indicated by the dotted lines at H. Now apply the gauge to the new staff, as shown in Fig. 10 and the taper end will locate the exact position for the balance seat.

As previously stated, I have taken it for granted that you preferred to finish all the lower portion of the staff while the work was held in the chuck. I have assumed that you prefer to work in this way because I have noted the fact that nine watchmakers out of every ten start with and first finish up, the lower portion of the staff. Where this method of working originated I do not know, but it always has the appearance to me of "placing the cart before the horse." I do not pretend to say that a true staff cannot be made in this way, but it certainly is not the most convenient nor advisable. We all know that the heaviest part of the staff is from the roller seat to the end of the top pivot. Now it seems to me that it is the most natural thing in the world for a mechanic to desire to turn the greater bulk of his work before reversing it. Now if the workman has been educated to turn indifferently with right or left hand, it may make little difference, as far as the actual turning is concerned whether he starts to work at the upper or lower end of the staff, but unfortunately there are few among us who are so skilled as to use the graver with equal facility with either hand; and it is therefore an advantage to start with the upper end, as you can thus finish a greater portion of the work more readily. You can readily see that when you come to reverse your staff and use the wax chuck, that by starting at the top of staff your wax has a much larger surface of metal to cling to, and again the shape of the balance seat is such as to secure the work firmly in the wax, while if the reverse method is employed, the larger portion of the balance seat is exposed and the staff is more liable to loosen from the motion of the lathe and pressure of the graver and polishers.

CHAPTER IV.

BY the aid of the pinion calipers and the old staff, the diameter of the roller seat and the balance and hair-spring collet seats may be readily taken, but it is perhaps better to gauge the holes, as the old staff may not have been perfect in this respect. A round broach will answer admirably for this purpose, and the size may be taken from the broach by means of the calipers. In fitting our pivots, we cannot be too exact; and as yet no instrument has been placed upon the market for this purpose which is moderate in price and yet thoroughly reliable. The majority of watchmakers use what is termed the pivot-gauge, a neat little instrument which accompanies the Jacot lathe, and which may be obtained from any material house. This tool, which is shown in Fig. 11, is, however, open to one objection in the measurement of pivots, and that is, that it may be pressed down at one time with greater force than at another, and consequently will show a variation in two measurements of the same pivot. Some of my readers may think that I am over-particular on this point, and that the difference in measurement on two occasions is too trivial to be worthy of attention, but I do not think that too much care can be bestowed upon this part of the work, and neglect in this particular is, I think, the cause of poor performance in many otherwise good timepieces. The ordinarily accepted rule among watchmakers is that a pivot should be made $\frac{1}{2500}$ of an inch smaller than the hole in the jewel, to allow for the proper lubrication. I am acquainted with watchmakers, and men who are termed good workmen, tqo, who invariably allow $\frac{1}{2500}$ of an inch side shake, no matter whether the pivot is $\frac{12}{2500}$ or $\frac{16}{2500}$ of an inch in diameter. Now if $\frac{1}{2500}$ of an inch is the proper side shake for a pivot measuring $\frac{12}{2500}$ of an inch in diameter, it is certainly not sufficient for a pivot which is one third larger.

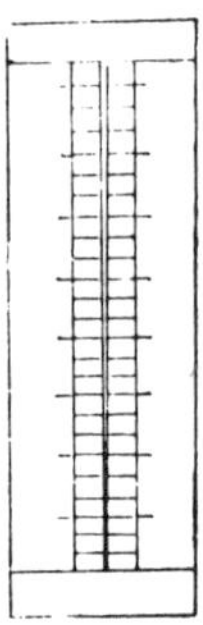

Fig. 11. Fig. 12.

Of course it is understood that side shakes do not increase in proportion according as the pivot increases in size, for if they did a six-inch shaft would require at this rate a side shake of ½ inch, or ¼ inch on each side, which would be ridiculously out of all proportion, as the $\frac{1}{64}$ of an inch would be ample under any circumstances. Neither can we arrive at the proper end shake for a pivot by reducing in proportion from the end shake allowed on a six-inch shaft, because if we followed out the same course of reasoning we would arrive at a point where a pivot measuring $\frac{12}{2500}$ of an inch would require an end shake so infinitely small that it would require six figures to express the denominator of the fraction. and the most minute measuring instrument yet invented would be incapable of recording the measurement. We must leave sufficient side shake, however, on the smallest pivot and jewel for the globules of the oil to move freely, and experiments have shown conclusively that $\frac{1}{2500}$ of an inch or $\frac{1}{5000}$ on each side of the pivot, is as little space as it is desirable to leave for that purpose, as the globules of the best chronometer oil will refuse to enter spaces that are very much more minute. But to return to our pivot guage.

Each division on the gauge represents $\frac{1}{2500}$ of an inch, which is all that we require. The diameter that the pivot should be, can be ascertained by inserting a round pivot broach into the jewel and taking the measurement with the pivot gauge, and then making the necessary deduction for side shake. Slip the jewel on the broach as far as it will go, as shown in Fig. 12, and then with the pivot gauge take the size of the broach, as close up to the jewel as you can measure, and the taper of the broach will be about right for the side shake of the pivot. If, however, you prefer to make the measurement still more accurate, you can do so by dipping the broach into rouge before slipping on the jewel and then remove the jewel and the place which is occupied on the broach can be plainly discerned and the exact measurement taken and an allowance of $\frac{1}{2500}$ of an inch made for the side shake. Another method, and one which is particularly applicable to Swiss watches, where the jewel is burnished into the cock or plate, is to first slip on to the broach a small flat piece of cork and as the broach enters the jewel the cork is forced farther on to the broach, and when the jewel is removed it marks the place on the broach which its inner side occupied, and the measurement can then be taken with the gauge. If care is used in the selection of a broach, that it be as nearly perfect in round and taper as possible, by a little experiment you can soon ascertain just what part of the length of the broach corresponds to one degree on the gauge and by a repetition of the experiment the broach can then be divided accurately, by very minute rings turned with a fine-pointed graver, into sections, each representing one degree, or $\frac{1}{2500}$ of an inch, and the measurement will thus be simplified greatly.

As before stated, much depends upon the condition of your gravers and the manner of using them. It is of the utmost importance that they be kept sharp, and as soon as they begin to show the slightest sign of losing their keenness, you should sharpen them. The proper shape for balance pivots was shown in Fig. 4. Now let us examine into the best positions for holding the gravers. In Fig. 13 two ways of holding the graver are shown, A representing the right and B representing the wrong way. If the graver is applied to the work as shown at A, it will cut a clean shaving, while if applied as shown at B it will simply scrape the side of the pivot and ruin the point of the graver without materially forwarding the work. Again, the holding of the graver is indicated at A has its advantages, because the force of the cut is towards the hand holding it, and should it catch from any cause the jar of the obstruction will be conveyed immediately to the hand, and it will naturally give and no harm

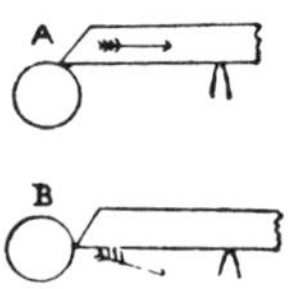

Fig. 13.

will be done. If, on the other hand, the graver should meet with an obstruction while held in the position indicated at B, the force of the cut will be in the direction of the arrow, downward and toward the rest, and the rest being unlike the hand, or rather being rigid, it cannot give, and the result is that the work, or graver, or both, are ruined. In Fig. 14 two other methods of holding the graver are shown. The general roughing out of a staff should be done with the graver held about as shown at A, Fig. 13; but in finishing, the graver should be held so that the cut is made diagonally, as indicated at A, Fig. 14. It is rather difficult to explain in print just how the graver should be held, but a little experiment will suffice to teach the proper position. The best indication that a graver is doing its work properly, is the fact that the chips come away in long spiral coils. Aim to see how light a cut you can make rather than how heavy. Never use force in removing the material, but depend entirely upon the keenness of the cutting edges. Never use the point of the graver, except where you are compelled to, but rather use the right or left hand cutting edges. By following out this rule you will find that your work, when left by the graver, requires little or no finishing up, except at the pivots. At B, Fig. 14, is shown the correct manner of applying the graver when turning a pivot. Hold the

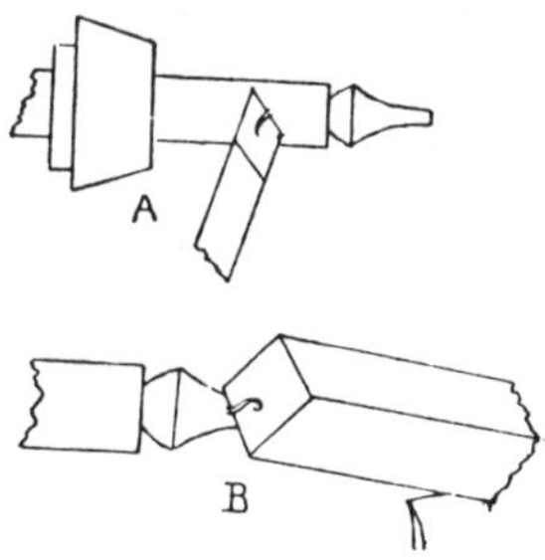

Fig. 14.

graver nearly on a line with the axis of the lathe and catching a chip at the extreme end of the pivot with the back edge of the graver, push slightly forward and at the same time roll the graver towards you and it will give the pivot the desired conical form. By keeping the graver on a line with the length of the pivot, all the force applied is simply exerted in the direction of the chuck, and does not tend to spring the pivot, as it would were the extreme point applied, as in Fig. 13. When we come to such places as the shoulder of the back slope, the seat for the roller, balance, etc., we must necessarily use the point of the graver.

CHAPTER V.

IN chapter IV I called attention to the right and wrong way of holding the graver while using the extreme point, and also the correct manner of applying the graver in turning conical pivots. I also called attention to the fact that it was well to only use the point of the graver where positively necessary, as in the back slope of the pivot, etc. In turning the seat for the balance, as

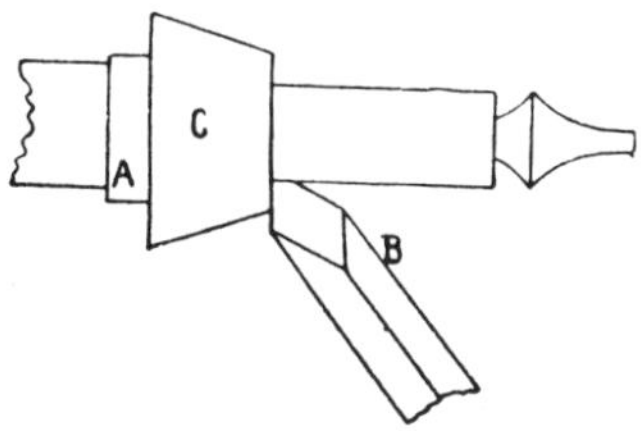

Fig. 15.

indicated at A, Fig. 15, the graver A, Fig. 1, or a similar one as shown at B, Fig. 15, should be used. The slope at C should now be turned. In turning the pivot and seat for the roller, you should leave them slightly larger than required, to allow for the grinding and polishing which is to follow. No definite amount can be left for this purpose, because the amount left for polishing depends entirely on how smoothly your turning has been done. If it has been done indifferently, you may have to allow considerable for grinding and polishing before all the graver marks are removed, while, on the contrary, if the work has been performed with care, very little will have to be removed. Avoid the use of the pivot file by performing your work properly to start with.

For grinding, bell-metal or soft iron slips are desirable, and the grinding is effected by means of oil stone powder and oil. Two slips of metal similar in shape to A and B, Fig. 16, are easily made, and will be found very useful.

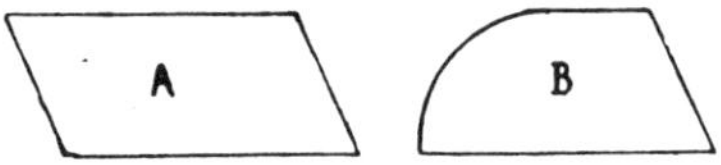

Fig. 16.

A is for square pivots, etc., while B is used for conical pivots. These slips should be dressed with a dead smooth file, the filing to be done crosswise, to hold the oil stone powder and oil. During the operation of grinding, the lathe should be run at a high speed and the slips applied to the work lightly, squarely and carefully. The polishing is effected by means of diamantine and alcohol. After the work is brought to a smooth gray surface, slips of box-wood of the shape shown in Fig. 16 should be substituted for the metal slips. Oil stone slips are sometimes used in lieu of metal ones, but they soon get out of shape and are troublesome to care for on this account. All things considered, there is nothing better for polishing than a slip or file made of agate, say one inch long, one-quarter inch wide and one-eighth inch thick. A slip of this kind can be obtained from any lapidary, and after grinding with emery and water until the surface has a very fine grain, it should be mounted by fastening with cement into a brass socket and this is then inserted into a small wooden handle, as shown in Fig. 17. The agate slip should be ground to about the shape of B, Fig. 16, so that one side can be used for square corners and the other for conical pivots. The final polish can soon be imparted by means of a small boxwood slip, or flattened peg-wood, and diamantine and alcohol. Never try to bring out the final polish until you are satisfied that all graver marks have been ground out, otherwise you will simply have to go all over the work again.

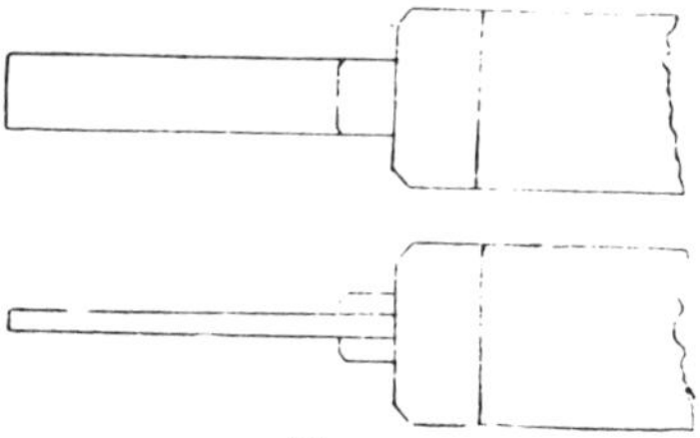

Fig. 17.

When the staff is finished from the lower pivot to the seat of the balance, the upper part should be roughed out nearly to size and then cut off preparatory to finishing the top part.

Attention was previously called to the fact that the majority of watchmakers prefer to finish all the lower portion of the staff first, notwithstanding the fact that there are numerous advantages to be gained by proceeding to first finish up the upper portion. We have now reached the point where the wax chuck must be used, and perhaps these advantages may be now more clearly defined. In order that the two procedures may be more distinctly shown, illustrations of both methods are here given. Fig. 18 shows the popular method, the lower portion of the staff being all completed and fastened by means of wax, in the wax chuck. Fig. 19 shows the opposite course of procedure. In both illustrations the lines indicate the amount of wax applied to hold the work. It will be noted that in Fig. 18 the hub of the staff is

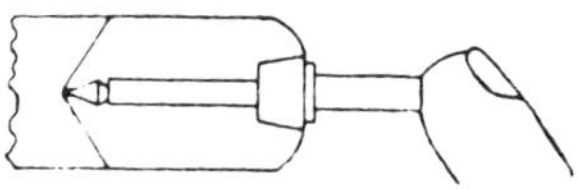

Fig. 18.

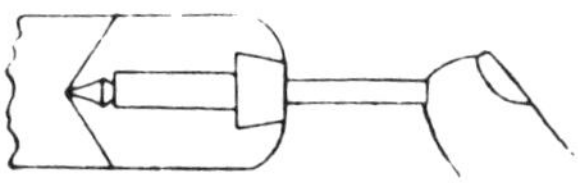

Fig. 19.

enclosed in the wax very much as a cork is fitted into a bottle, while in Fig. 19 the hub is reversed, just as a cork would appear where the larger portion within the bottle and the smaller portion protruding through the neck. A study of the diagram will readily show that in Fig. 19 the staff is held more rigidly in place and that a greater bulk of the work is enclosed in the wax than in Fig. 18, although there is less wax used in the former than in the latter.

Before proceeding to set the staff in the wax, it is necessary to make some measurements to determine its full length. Remove both cap jewels and screw the balance cock in place. Examine the cock and see if it has at any time been bent up or down or punched to raise or lower it. If so, rectify the error by straightening it and then put it in place. Now with a degree guage, or calipers, proceed to take the distance between the outer surfaces of the hole jewels and shorten the staff to the required length. Do not remove too much, but leave the staff a little long rather than cut it too short, as the length can be shortened later.

A very handy tool for the purpose of making these length measurements can be constructed by adding a stop screw to the common double calipers as shown in Fig. 20. The improvement consists in the fact that they can be opened to remove from the work and closed again at

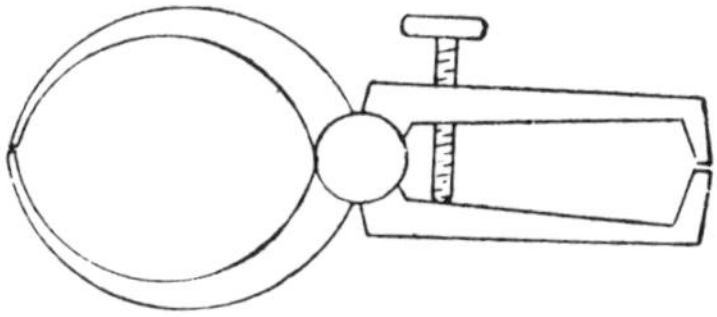

Fig. 20.

exactly the same place, so that an accurate measurement can be made. The all-important point in the use of wax chucks is to get a perfect center. If you are not careful you are liable to leave a small projection in the center as shown at A, Fig. 21. The ordinary wax chuck cannot be unscrewed from the spindle and restored to its proper place again with anything like a certainty of its being exactly true, and if you insist on doing this there is no remedy left but finding a new center each time. If will be found more satisfactory and economical in the long run to have a permanent chuck for a wax chuck and you will then have no necessity for removing the brass chuck.

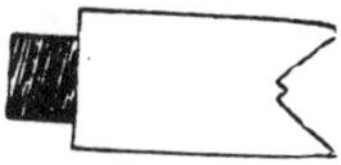

Fig. 21.

The center, or cone for the reception of the pivot, should be turned out with the graver at an angle of about 60°

and such a graver as is shown at B, Fig. 1, will answer admirably for this purpose. After you have carefully centered your wax chuck, place a small alcohol lamp under the chuck and heat it until the wax will just become fluid and yet not be hot enough to burn the wax. Revolve the lathe slowly and insert the staff so that the pivot rests squarely and firmly in the center. Now re-heat the chuck carefully in order that the wax may adhere firmly to the staff, keeping the lathe revolving meanwhile, but not so fast that the wax will he drawn from the center, and at the same time apply the forefinger to the end of the staff, as shown in Figs. 18 and 19, and gently press it squarely into place in the wax chuck. The lines in Figs. 18 and 19 designate about the right amount of wax after the work is ready, but it is well to add a little more than is shown in those figures, and you should be careful to keep the wax of equal bulk all around, or when it cools it will have a tendency to draw the staff to one side. Now remove the lamp and keep the lathe revolving until the wax is quite cool, when it should be removed, by means of a graver, down to the dimensions designated by the lines in Figs. 18 and 19. When this is accomplished re-heat a little, but only enough to make it soft, but not liquid, and placing a sharpened peg-wood on the tool rest proceed to the final truing up, by resting the pointed end against the hub.

CHAPTER VI.

I HAVE described above one of the methods in vogue for holding a staff by means of wax. It is the common method employed by most watch repairers, the popular method, so to speak. The method which I am now about to describe may seem awkward at first to those who have not practiced it, but once you have fairly tried it, you will never be contented to work in any other way.

The first requisite is a true taper chuck; and it is well to purchase an extra one to be used solely for this purpose, so that you will be prepared at all times for staff work. Select a good steel taper, and having placed your chuck in the lathe, see if your taper fits well by inserting it in the chuck while running slowly. If it fits well, it will be marked almost throughout its length. Insert again in the chuck, and with a few light taps of the hammer set it firmly in place, so that you know that there is no danger of its working loose. The taper will then project about three-quarters of an inch from the face of the chuck. By means of a sharp graver, make the face of the taper smooth and straight, and cut off the taper end. Now mark a point on the taper about one-fourth of an inch from the end and proceed to turn down the diameter from this point to the end, leaving that portion of the taper about two-thirds of its original diameter, and finish with a nice square shoulder. Now with a long-pointed sharp graver proceed to cut a nice V-shaped center with an angle of about 60°. When you have proceeded thus far you will find that you have an implement resembling that shown in Fig. 22.

Care must be taken that the center is quite true, and that no projection is left like that illustrated in Fig. 21, no matter how minute it may be. Now examine the center by the aid of a strong glass, and after you are satisfied with its appearance proceed to test it. Take a large

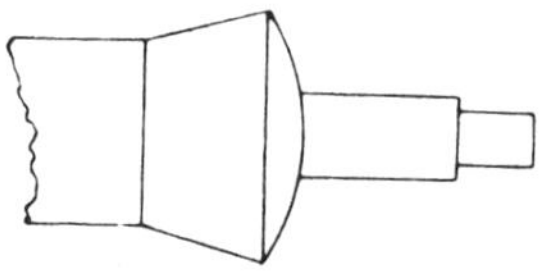

Fig. 22.

sized pin with a good point, and placing the point in the center, maintain it in position by pressing upon the head, and while revolving the lathe slowly proceed to examine by means of your glass. If the center is a good one there will be no perceptible vibration of the pin.

Now procure a piece of small brass tubing with an internal diameter a little less than that of the turned down portion of your taper. If the brass tubing cannot be procured readily, you can substitute a piece of brass wire a little larger than the taper, and by means of a drill a little smaller in diameter than the turned down portion you can readily make a small tube about one-half inch long. Now by means of a broach proceed to open the tube to a point one quarter inch from one end, and carefully fit it on the turned down portion of your taper. After fitting tightly to the shoulder of the taper, proceed to turn out the other end until it will take in the hub of your staff easily and leave a little room to spare. Now turn your tube down in length until a little of the hub is exposed either way you put the staff in. Turn the outside of the tube smooth and to correspond with the outline of the taper, so you will have a nice looking job when completed. Just below where the hub will come drill a small hole in the tube and remove all burr, both inside and out, that may have been made in drilling, so that the shellac or wax will not adhere to it. This little hole acts as an outlet for the air in the tube; and as the hot shellac enters at the end of the tube the air is expelled through this vent. It also helps to hold the cement firmly in place. Now try your staff in the tube again, and be sure that it is quite free, and that you will be able to work on the portions of it above and below the hub, according as one end or the other is inserted.

You are now ready to insert your staff and proceed with your work. Hold your shellac in the flame of your lamp a moment until it is quite liquid, and then smear both the inside and outside of the tube with it. Heat the shell or tube gently by means of the lamp, keeping the lathe revolving slowly all the while, and taking the staff in your tweezers proceed to insert it carefully into the tube. Press firmly back, making sure that it has reached the bottom of the V-shaped center. Pack the cement well in around the staff, and while centering remove the lamp and allow the whole to cool, keeping the whole revolving until quite cool. Now remove the superfluous cement by means of the graver, and heating tube again slightly, proceed to center exactly by means of a pointed peg-wood, resting on your T rest to steady it. Turn slowly in the lathe and examine with glass to see that it is quite true. Your completed instrument will resemble Fig. 23.

The advantage of the device is that your center is always ready, and all you have to do is to insert your chuck in the lathe, warm it, and you are ready to insert your staff and proceed to work. As I said in the first

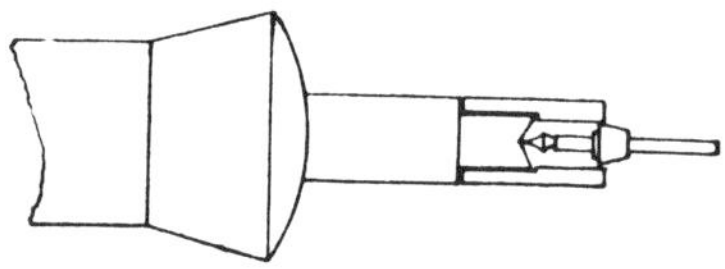

Fig. 23.

place, it is well to employ a taper chuck exclusively for this work, and not attempt to use it for any other, for if you try to remove your taper and replace it again, you will surely find that your work is out of center, and you will be compelled to remove the brass shell and find a new center each time you use it. You can avoid all this trouble, however, by purchasing an extra chuck and devoting it exclusively to wax work. Of course, the brass shell can be removed and placed in position again without in any way affecting the truth of the center, and any number, shape and size of shells can be made to fit the one taper, and these shells will be found very useful for holding a variety of work, aside from balance staffs.

CHAPTER VII.

THE two popular methods of holding a balance staff in wax have been described and illustrated; the reader may take his choice. The turning and finishing of the other end of the staff is performed as previously described. That portion on which the hair-spring collet goes should be turned to nearly the proper size, making due allowance for the grinding and polishing that is to come. The balance seat should be slightly undercut, so that the balance can be driven on tightly and all riveting dispensed with. The size for the pivot can be determined from its jewel, as previously described. Finish the ends of the pivots flat and round the corners off slightly; and right here comes a point worthy of consideration in all watch work. Leave no absolutely square corners in any of your work, but round them off very slightly. This may seem a very little thing, but it is one of the small things that go to make up first-class work. You can judge pretty accurately of a watchmaker by the corners he leaves on his work, as well as by the appearance of his gravers and screw-drivers.

When your staff is completed and nicely polished, remove from the wax and boil in alcohol to clean, and when dried it is ready for the balance. Great care must be exercised in removing the balance from the old staff, especially if it be a compensation balance, that you do not distort it any way. If the balance has been riveted on extra care will have to be exercised. The riveting may be cut by means of a graver, or a hollow drill made from Stubb's steel wire. The recess in the drill should just fit over the shoulder left for the reception of the hairspring collet. The edge of the hollow drill has small teeth formed upon it similar to a fine file, and will cut quite rapidly.

After removing the balance, if it appears to be sprung in the arms, the result of removal or previous bad treatment, proceed to bend them straight, and then to true up the rim carefully, and stake on with a flat end punch. Now put on your roller and drive it down to the hub and see that the roller is free from the fork. See that jewel pin reaches fork properly and that the guard pin also reaches the roller. See that your balance is free from the plate and the bridge. If the balance is true and all right, you are ready to put on your hairspring. See that it is in beat. It is well to make a mark on the balance before taking off the old staff, showing positions of hairspring stud and jewel pin.

Three-quarter plate English lever and Swiss lever balance staffs differ only in detail, except that they are sprung under balances. The general operations for making, however, are similar to those described.

I have not described the method of poising the balance for two reasons; first, the mere poising of a balance for a cheap movement is so simple that it needs no explanation; and second, to describe the poising of the balance of a fine watch is a lengthy task, and can hardly be included under the heading of staffing and pivoting. The ground has been thoroughly and conscientiously covered by Mr. J. L. Finn, in a little volume entitled Poising the Balance,* and I would advise all watchmakers, both young and old, to read what he has to say.

Good pivoting is an art in itself, and although there are many who undertake to do this work, there are but few who can pivot a staff in such a manner that it will bear close inspection under the glass. We often hear watchmakers brag of the secrets they possess for hardening pivot drills, but I fancy they would be somewhat surprised if they traveled around a little, to find how many watchmakers harden their drills in exactly the same way that they do. The great secret, so-called, of making good drills, is to first secure good steel, and then use care to see that you do not burn it in the subsequent operations. The fewer times the steel is heated the better. My experience teaches me that you can do no better than to select some nice pieces of Stubb's steel for your pivot drills. Many watchmakers make their drills from sewing needles, say No. 3 or 4, sharps. The steel in these needles is usually of good quality, but the great drawback is that a drill made from a needle will not resist any great pressure, and is liable to break just at the time

---

*POISING THE BALANCE, by J. L. Finn, Geo. K. Hazlitt & Co., publishers Chicago.

that you have arrived at the most important point. If your drill is made from a piece of Stubb's steel wire, or an old French or Swiss graver, you not only know that the material in it is first-class, but you can leave the base of the drill solid and substantial, with enough metal in it to resist considerable pressure. The part of the drill which actually enters the pivot is very short, and the end can be turned down to the desired diameter. Turn or reduce your wire by means of a pivot file so as to be smooth and conical, as shown at A, Fig. 24. The conical form is given to the drill for exactly the same reason that it is given to the balance pivots, because it gives additional strength. Heat to a very pale red for about one-half inch from the end, and then spread the point, as shown at B, Fig. 24, by a slight blow of the hammer. We are now ready to temper our drill, and we must exercise a little care that the steel is not burnt and that the drill is not bent or warped when hardening. The flame of the alcohol lamp should be reduced as small as possible, or otherwise the steel may become overheated and lose all its good qualities. If needles are used for making drills there is a great liability of their warping when hardening, but when a larger piece of wire is used there is not much danger, if care is exercised in introducing the drill that it goes into the compound straight and point foremost. If a needle is used, it is well to construct a shield for it, to

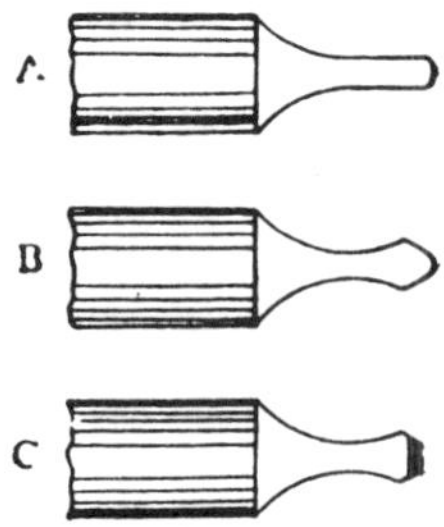

Fig. 24.

be used when heating and hardening. This shield can be made from a small piece of metal tubing, broached out to fit loosely over the shank and point of the drill. The drill is introduced into this shield as shown in Fig. 25, and a little soap may be introduced into the end a before plunging. Various hardening devices are used, but in my experience beeswax or sealing wax will be found as good as any. Heat the drill (or if a needle, the drill and shield both), to a pale red and plunge straight into the wax. In the latter case, where the shield is used, the shield, on striking the wax, will run up the shank of the drill, allowing the point to pierce the wax. Some watchmakers introduce the extreme point of the drill into mercury first and then plunge into the wax. This hardens the extreme point of the drill very hard, so hard, in fact, that it will penetrate the hardest steel, but care must be exercised with such a drill because the mercury makes it not only very hard but very brittle. C, Fig. 24, shows a drill after it has been finished on the Arkansas stone. This shape of drill will withstand the pressure necessary to drill into hard steel. Many watchmakers reduce the temper of every staff before drilling. This, I think, is quite unnecessary. There are very few cases in which it is necessary to reduce the temper of the staff, and even then it should only be reduced as far as it is to be drilled, and then not in excess of a good spring temper.

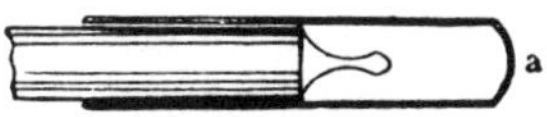

Fig. 25.

The centering of a staff in wax has been thoroughly described and, in pivoting, the proceeding is the same as in staffing. After accurately centering your work, make a small cut in the center for the reception of the drill and make this mark deep enough to take the entire cutting head of the drill. Keep the drill firmly pressed into this center and keep wet constantly with turpentine. Do not revolve the work all one way, but give the lathe an alternating motion. At first give but a third or a half revolution each way, until the drill begins to bite into the staff, when you can then safely give it a full revolution each way. Care must be exercised, however, not to give the work too rapid a motion, for if you do the friction is apt to draw down the temper of your drill. Many watchmakers find that their drills cut well for a certain distance and then refuse to work altogether, and one of the chief reasons is that they are in too great a hurry with their drilling.

If you find it absolutely necessary to reduce the hardness of your staff before drilling, do so by drilling a hole in the end of a small piece of copper wire that will just fit over the part to be softened, and apply the heat to this copper wire, say one-fourth of an inch from the staff. The heat will run down the copper wire and heat the staff just where you wish to draw the temper. Be careful and do not draw the temper too much, nor let it extend down the staff too far.

The plug for the new pivot should be carefully made, perfectly round, with a very little taper, and should be draw-filed before being driven in. Some workmen dip the plug in acid before driving in, as they declare that the pivot is less liable to be loosened while turning, if so treated. The acid simply rusts the pivot and the hole, but I cannot see that this will hold it any more firmly in place while finishing. If the taper is a gradual one and the pivot a good close fit, there will be little danger of it loosening while dressing to shape. If too great a taper is given to the plug, there is danger of splitting the end of the staff, and this involves the making of an entire new staff.

The turning up of a new pivot does not differ in any way from the instructions given for turning pivots on a new staff. With a little care both in turning and finishing, a new pivot can be put in so nicely that only the initiated can tell it, and then only with the aid of a strong glass.

In pivoting cylinders there is some danger of breaking them. To avoid this, select a piece of joint wire, the opening of which is slightly larger than the diameter of the cylinder at the lower end, and cut off a piece the length of the cylinder proper, leaving the pivot projecting. Now fill the cylinder with lathe wax, and while the wax is warm, slip on the joint wire. You can now proceed to true up the pivot in the usual manner, and when the wax is quite cold, proceed to turn and polish the pivot before removing from the lathe. If the joint wire is properly cemented on the cylinder, it is almost impossible to break it. After all the work is done, the wax can be dissolved in alcohol. In pivoting pinions to cylinder escape-wheels and third wheels, it is not necessary to remove the wheels, but great care should be used in handling. In the latter case use plenty of wax. Do all your centering by the outside of the pinion. Perfect centering and sharp tools are requisite to good pivoting. Do not try to rush your work, especially while drilling. Proceed deliberately with your work and aim to restore the watch to the condition it was in originally, and you will find staffing and pivoting is not half as hard as some workmen would have you believe.

CLOCK WHEEL & PINION CUTTING

by **J. Malcolm Wild**

This little book comprehensively covers how to calculate and cut wheels and pinions on a conventional lathe and is written by an expert in professional clock repair.

48 pages - 5 ½ x 8 ½ Perfectbound...$**11.95**

Available again!

The MODERN CLOCK

by **Ward Goodrich**

A Study of Time Keeping Mechanism; Its Construction, Regulating and Repair (original published in 1905).
This is one of the best book on the market for repairing American and European clocks; including Cuckoo, Electric and 400-Day clocks.

512 pages - 5 ½ x 8 ½ - Perfect bound ...... $**19.95**

WHEELS & PINIONS

& How to Determine Their Exact Size
by **H. Shouffelberger & T. Gribi**

This book contains nine tables and one plate which are all that is necessary for the watchmaker to determine the exact size of missing wheels or pinions.
The tables apply to both systems of measurement i.e. the English inch or metric system and are more convenient and faster to use than the old method.
20 pages - 8 ½ x 11 .... $**5.00**

STAKING TOOLS

AND HOW TO USE THEM

By K & D Manufacturing Corp.

Contains a brief history of the development and use of the Staking Tools set. It also contains the first several pages of the 1910 Catalog of Staking Tools and constitutes a hand-book every clockmaker needs on his bench on the various jobs related to the use of the Staking Tool.

108 pages - 5 ½ x 8 ½ - Perfectbound **$11.95**

PRECISION TIME MEASURES

by **Charles T. Higginbotham**

Contains: The Mainspring. Calculation of Watch Trains. Watch Jewels. Escapements. The Detached Lever. Making A Drawing of An Escapement. The Cylinder Escapement. The Balance Staff. Making the Balance. Cleaning a Watch. Balance Truing. Poising the Balance. Springing. Magnetism. Adjusting and useful tips for workmanship.

343pp - 5½ x 8 ½ - Hardbound **$19.95**

Helpful Information for Watchmakers

By Olof Ohlson

Contains: Time & Timing. Waltham Lever Escapement. Pallet Action. Roller & Jewel Pin. Matching the Escapement. Jewel Pin Action. Safety Action - Directions for putting the Escapement in Beat. Mainsprings. Mainspring Don'ts. Table showing Thickness Strength of Mainsprings. Table Showing Width of Mainsprings. Mainsprings. A Proper Method for Cleaning a Watch Movement. How to Set a Jewel in a Watch Plate. How to Remove a broken Screw from a Watch Plate. Casing. Some Hints on Measuring as applied to Watchmaking. Taper Shoulder & Detachable Balance Staff. Hairspring Stud. Spiral Winding Pinion & Crown Wheel. Jeweled Main Wheel. Taper Steady Pins. Pendant Setting Mechanism. Waltham Hairsprings. Hints & Suggestions Waltham Screw Taps. The Size of a Watch Movement - Useful Numbers. Table of Equivalents & Measurements of Lengths etc...

104 pages - 5 ½ x 8 ½ - Perfectbound.. **$11.95**